Mustapha Chelghoum

Plant histology

Mustapha Chelghoum

Plant histology

Abstract

Imprint

Any brand names and product names mentioned in this book are subject to trademark, brand or patent protection and are trademarks or registered trademarks of their respective holders. The use of brand names, product names, common names, trade names, product descriptions etc. even without a particular marking in this work is in no way to be construed to mean that such names may be regarded as unrestricted in respect of trademark and brand protection legislation and could thus be used by anyone.

Cover image: www.ingimage.com

This book is a translation from the original published under ISBN 978-620-6-71194-0.

Publisher:
Sciencia Scripts
is a trademark of
Dodo Books Indian Ocean Ltd. and OmniScriptum S.R.L publishing group

120 High Road, East Finchley, London, N2 9ED, United Kingdom
Str. Armeneasca 28/1, office 1, Chisinau MD-2012, Republic of Moldova, Europe
Printed at: see last page
ISBN: 978-620-7-62020-3

PLANT HISTOLOGY

ABSTRACT

CHELGHOUM MUSTAPHA

FOREWORD AND OBJECTIVES

The vascular plants that colonise the land are the most important source of various human needs, such as food, protection and care. The study of plants has been a necessity since the dawn of time. The need to classify plants so that they can be identified has led to the development of various techniques for studying plants. The morphology of a plant is the consequence of the characteristics of the internal structure of its organism. The development of microscopy has made it possible to study this internal structure. A plant organism is made up of a set of cells organised into tissues. The study of tissues is called histology. Biologists use histology to understand other fields such as physiology and biochemistry. Plant histology is particularly important in pharmacognosy, where drugs are made up of plant fragments whose external morphology cannot be used to identify them. Certain histological characteristics allow species to be classified even at lower levels.nThis handout is dedicated to the description of plant tissues. It is divided into three chapters: the first deals with the plant cell, the second describes the different plant tissues, and the third describes the organisation of tissues in the different plant organs, known as plant anatomy. Simple and sufficient descriptions have been accompanied by pictorial illustrations, making it easy to understand the different concepts.

TABLE OF CONTENTS

CHAPTER I

THE VEGETABLE CELLULE

1. Definition

The cell is the structural and functional unit of the living organism. Plants are made up of eukaryotic cells (true nuclei bounded by a membrane), which differ from animal cells in a number of ways that have important implications for the structure and function of the entire organism. Cells vary in size, shape, structure and function.

2. Structure of the plant cell

The plant cell consists of a droplet of living matter: protoplasm, compartmentalised by membranes and surrounded by a rigid wall (Figure 1). The cell is made up of 90% water and 10% dry matter.

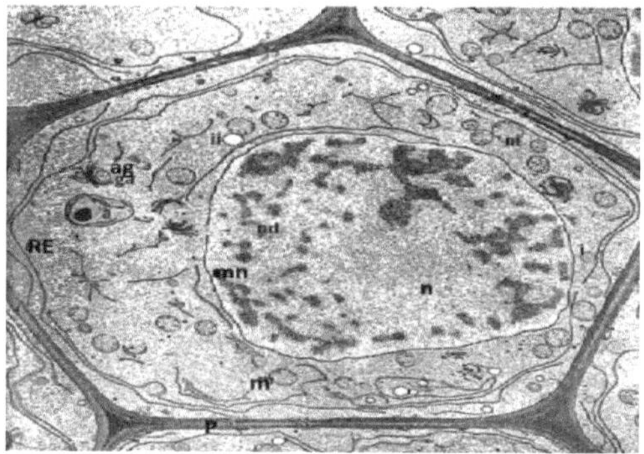

Figure 1: Ultra structure of a meristematic cell

Transmission electron micrograph of a meristematic cell in the root cap of Zea mays (maize). The protoplast, dominated by the nucleus (n), contains mitochondria (m), Golgi bodies (ag) and an endoplasmic reticulum (er). The protoplast is surrounded by a cellulose primary wall (P). Note that in this meristematic cell, only a few small (white) vacuoles are present. The dark structures in the nucleus are chromosomes. Magnification × 5973. Adapted from Whaley et al.

a. The protoplasm: made up of a base substance that is a transparent lipoprotein hydrogel (hyaloplasm) in which the cytoplasm is hollowed out. enclaves: hydrophilic (the vacuole) and hydrophobic (the lipid inclusions), and encapsulate the living elements: the nucleus, the chondriome, the plastidome, the endoplasmic reticulum and the dictyosome.

b. The vacuole: the vacuole contains various solutes at a concentration of 0.5 M. It also contains several hydrolytic enzymes, which help to differentiate certain cells by lysing the cell content, leaving only the cell wall. The vacuole may contain pigments and protein substances.

c. The nucleus: consists of a double porous membrane, the nuclear membrane, delimiting a nuclear juice (water, mineral salts) in which one or more nucleoli and chromatin are immersed. The nucleus gives the cell its living character and is responsible for hereditary transmission.

d. The chondrioma: made up of granulation, mitochondria, the site of cellular respiration, and filaments, the chondrioconte. In some cases, these granules are grouped in chains, the chondriomite.

e. The plastidome: composed of plastids. Different types exist. A plastid is delimited by a double membrane. The outer membrane is continuous, while the inner membrane has invaginations into the stroma. The stroma contains ribosomes and DNA, different from those in the nucleus. Lipid droplets and starch are also present. The precursor of all plastids is the proplast, which differentiates into the chloroplast, where photosynthesis takes place, the amyloplast, where starch is stored, and the chromoplast, where pigments accumulate.

f. The endoplasmic reticulum (ER): a system of saccules, some of which carry granules, the granular ER, the site of protein synthesis, and others which do not, the smooth ER, the site of transport of synthesised proteins.

g. The dictyosome or golgi apparatus: a collection of saccules and vesicles used to excrete proteinaceous material.

h. The membrane is made up of two phospholipid layers in which several proteins are inserted; the membrane that delimits the cytoplasm from the periphery of the cell is called the plasma membrane or plasmalemma, and is continuous from one cell to the next via cytoplasmic bridges in the lumen of the cells. plasmodesmata. The second type of membrane delimits the vacuole, the tonoplast. These membranes control the passage of substances to and from the cellular compartments.

i. The wall: pectocellulosic in nature, the wall is composed of three elements:

• A framework of cellulose, hemicellulose and peptic chains

• A complex cement of pectins and associated elements

• Water (up to 80% of the wall mass)

Transmission electron micrograph of a cross-section of parts of the walls of three adjoining cells of Pseudotsuga (Douglas fir) showing the middle lamella (ML), the primary walls (P) and the three layers of secondary walls (S1, S2, S3).

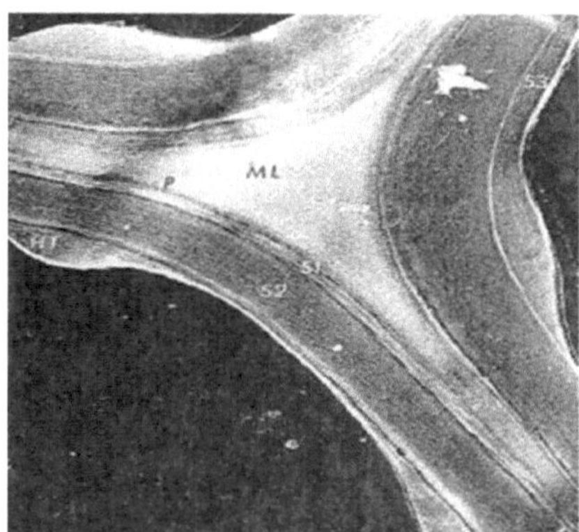

Figure 2: Ultra wall structure

The wall is made up firstly of the primary wall, which is the first and only wall formed in undifferentiated cells; it is capable of growing in length and thickness, and of the middle lamella shared by two neighbouring cells. The secondary wall is then formed in differentiated cells (Figure 2).

3. Meristem and meristematic cells

a. Structure of the meristematic cell

The tool without which the plant could not develop is a group of cells with particular characteristics, the meristematic cells. All the meristematic cells make up the meristem. A meristematic cell (Figure 3) can be characterised by its undifferentiated state. It is generally small and isodiametric, with little or no vacuolation. Its wall, devoid of tertiary incrustations, is pectocellulosic and not very thick. The cytoplasm is relatively dense and the nucleus voluminous. The organelles that accumulate there have not yet begun to differentiate, if at all. Their numbers are still low.

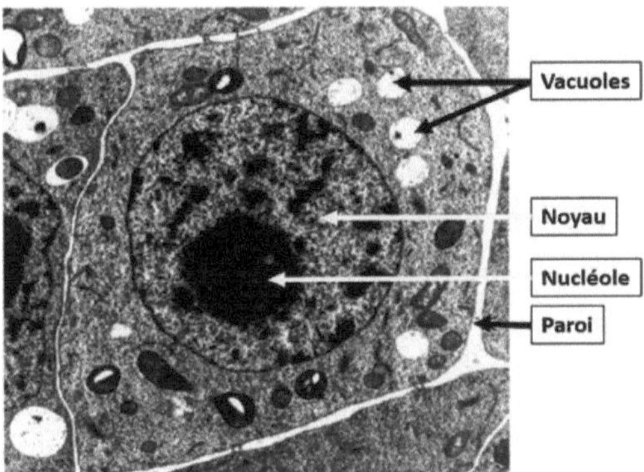

Figure 3: Structure of a meristematic cell

As well as these elements of undifferentiation, which mean that all the cells in a meristem are very similar, the meristematic state is accompanied by totipotency (the ability to differentiate according to the environment) and mitotic power in particular.

b. Location of meristems

A meristem will be active where it is born: this is a direct consequence of the impossibility of cell migration in plants. Its spatial location will therefore have a clear impact on the morphology of the organ it forms. Similarly, the structure it puts in place will depend on the precise moment at which it begins to function. Depending on whether it begins its life at the beginning or at the end of its development, it will give rise to very different formations. If we take into account these two factors, space and time, we can divide a plant's meristems into several categories.

i. Embryonic meristems

In addition to the two main meristems (root apical meristem and stem apical meristem), a number of meristematic cells are found at the base of the cotyledons, ensuring their development. The root apical meristem and the cauline apical meristem are always positioned at the ends of the embryo axis: these are the apical meristems.

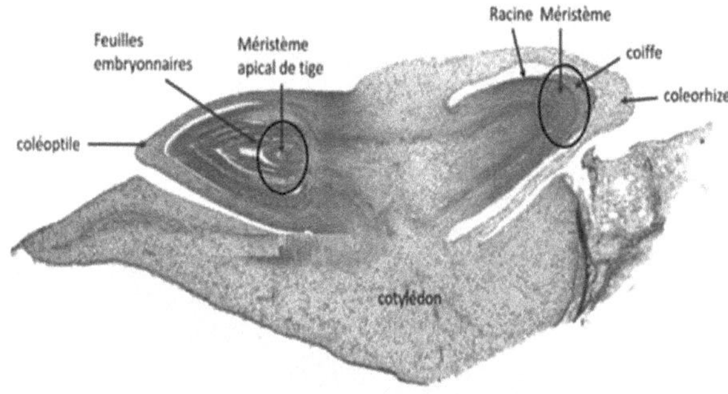

Figure 4: Embryonic meristem

ii. Primary meristems

• **In the stem:** First of all, there is a cauline apical meristem (Figure 5) located at the apex of the elongating main stem. This is in fact the meristem derived from the embryo which has continued to function in the stem. This meristem is joined by a number of lateral meristems which, when they become active, will be able to produce the plant's lateral branches. They are protected inside special structures developed by the stems: the lateral buds.

Figure 5: Structure of the apex of a vegetative shoot, note the pair of leaf primordia (1) at its base and a pair of bud primordia (1) in their axils. The tip of the apex (3) consists of a small, densely cytoplasmic cell, but the rib meristem cells are vacuolated and become the pith (4) of the young stem. The margin of the apex consists of densely stained margin meristem cells (5). Procambium (6)

• **In the root**: There is only one meristem: the root apical meristem (Figure 6), generally in a subapical position because it is always covered by the cap. However, the pericycle can be considered, at least partially, as a meristem. This is where the lateral roots of the underground organ are born.

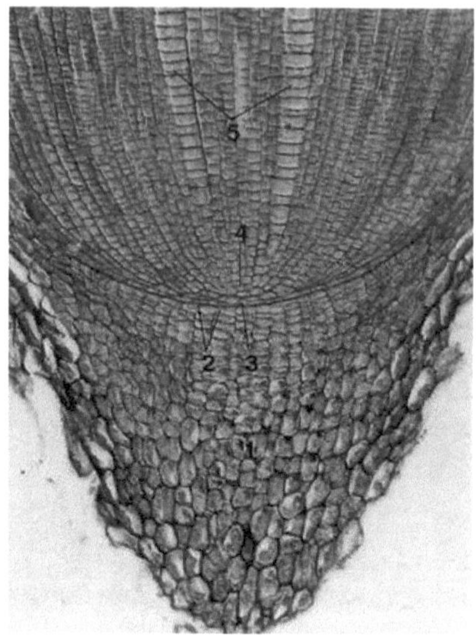

Figure 6: Structure of the root tip. The figure shows a prominent cap covering the apex. The cap (1) has its own distinct initials (calyprogen, 2) while the epidermis and cortex apparently arise from a common level of initials (3) adjacent to the calyprogen. The central cylinder of the procambium has its own initials (4). Note the rows (5) of hypertrophied cells in the procambium which represent the future conducting element.

• **In the leaf:** Knowledge of leaf meristems is more fragmentary than for the stem and root. It seems that, initially, a basal meristem initiates leaf growth. Later, marginal meristems located at the edge of the leaf blade take over.

iii. Secondary Meristems

In addition to the primary formations, there are secondary formations produced by secondary meristems called cambiums. Originally, a cambium appears as a cylinder of undifferentiated cells extending over all or part of the organ in which it is located. This cylinder generally consists of a cambial zone containing several layers of cells. The cells of the secondary meristem divide in a periclinal (tangential) mode, resulting in cells on one side (towards the outside) and others on the other (towards the inside) of the cambium, and in an anticlinal (radial) mode, which is essential for increasing the cambial circumference.

CHAPTER II

PLANT HISTOLOGY

1. Definition

A tissue is a group of cells with the same origin and the same main function. A tissue can be homogeneous when it is formed by a single type of cell with the same morphology, or heterogeneous formed by morphologically different cells.

2. Tissue classification

Fabrics can be classified according to their origin as :

• Primary: derived from primary meristems

• Secondary: from secondary meristems.

And according to their main function in :

• Covering tissues: organ protection

• Parenchymatous tissues: chlorophyll assimilation or reserve.

• Support tissues: reinforcing the organ's portability

• Conductive tissues: conduction of raw and processed sap

• Secretory apparatus: secretion of organic substances.

3. Tissue specialisation from thallophytes to cormophytes

The diversity of tissues that characterises vascular plants (tracheophytes) is not non-existent in thallophytes and bryophytes, but it is less marked. In thallophytes, for example, hyphae can agglomerate to form pseudo-tissues (prosenchyma and pseudo-parenchyma). Tissue specialisation becomes more marked in pteridophytes, which have an epidermis, parenchyma and conductive tissues. The primary meristem functions to form primary tissues, while the secondary meristem functions to form secondary tissues. These tissues are grouped according to their role in: tissues lining, parenchyma and support tissues, conductive tissues and secretory tissues.

4. Morphology of plant tissues

In addition to the function of the tissue, the morphology of its cells differs from one type to another. Tissues are studied using microscopy techniques, after sections have been cut from a plant organ. The sections are pre-treated with special dyes so that they can be differentiated under the microscope.

In practice, the following elements are used to identify a fabric:
- **Location in the organ**: on the surface of the organ, in the bark, at the limit of the central cylinder and the bark, in the bark or in the pith, etc.
- **Number of layers**: a tissue can be made up of one or more layers or a cluster of cells.
- **Cell junctions:** cells may be joined, separated by meats (triangular spaces between three adjacent cells) or gaps (wide spaces between several cells).
- **Cell shape and size**: cells can have various shapes: rectangular, sinuous, polygonal... small or large...
- **The nature of the wall**: primary or secondary; thick or thin; lignified or suberified, etc.
- **The state of the cell**: alive or dead

I. Plant tissues of primary origin

1. Covering fabrics

i. Definition: Protective tissues protect the body from the external environment, limiting water loss and forming a barrier against pathogens. They owe their properties to the impregnation of their cell wall by lipid derivatives (cutinisation, suberification), but they do not completely prevent gas exchange.

ii. Cauline covering tissue (The epidermis) :

This is a superficial tissue of primary origin consisting of a single layer of interlocking polygonal cells without meats between them, and oriented along the length of the organ. The inner and lateral walls are thin and cellulosic, while the outer prey is often thick and may be covered by a coating called the cuticle. The cuticle is made up of wax and cutin, substances of lipidic origin (Figure 1).

The epidermis may bear hairs resulting from the elongation of certain epidermal cells. These are tector hairs which play a protective role and have a cuticle like the epidermal cells. Their shape varies: single or multi-cell, single or multi-serial, massive or branched. Some hairs are made up of cells independent of the epidermis and have a secretory role. These are known as secretory hairs and form part of the secretory apparatus (Figure 1).

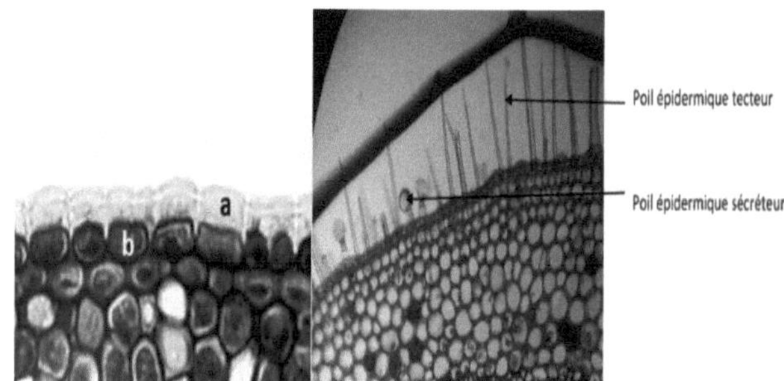

The epidermis: The outer periclinal walls of the epidermal cells are impregnated with lignin and cutin, and have expanded considerably. a: cuticle, b: epidermal cell

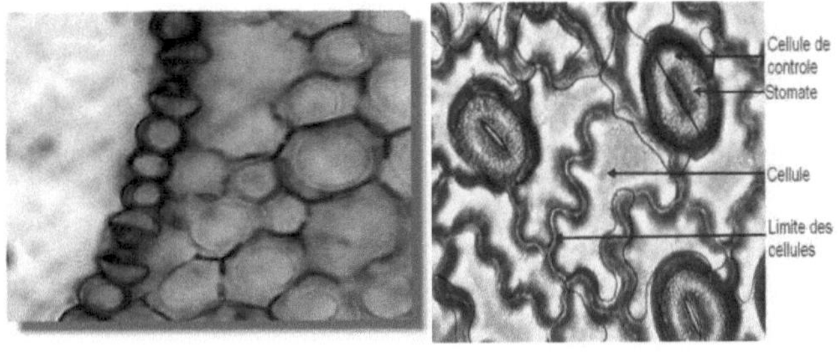

Cross-section through the epidermis, noting the stoma

Surface view of the epidermis of a leaf. Note the stomata

Figure 01: Epidermis

The epidermis is not a continuous tissue, but has openings in the epidermal base through which gas is exchanged with the atmosphere, known as aeriferous stomata, or from which water droplets exude (water-bearing stomata). A stoma is made up of two kidney-shaped cells, often chlorophyllous, with an opening between them, the ostiole; below the stomatal cells is a lacuna, the sub-stomatal chamber. The membrane of the stomatal cells is thick on the side of the ostiole and thin on the opposite side (Figure 1).

iii. Root linings :

Depending on the root zone, different tissues may be found. In the absorption zone, there are two root-covering tissues: the piliferous base, which also plays an absorption role, and the subereous base in dicotyledons, or the suberoid in monocotyledons.

The piliferous bed: a bed of cells, some of which have an extension forming an absorbent hair. The piliferous bed protects the absorption zone. These are living cells with a pectocellulose wall and a large turgid vacuole. They are destroyed upwards and regenerate downwards (Figure 2).

Suberified tissue : Above this region, the root is protected by tissue with a suberized wall. This is a single layer of cells called the suberous layer or several layers called the suberoid. These cells are contiguous and polygonal in shape (Figure 2).

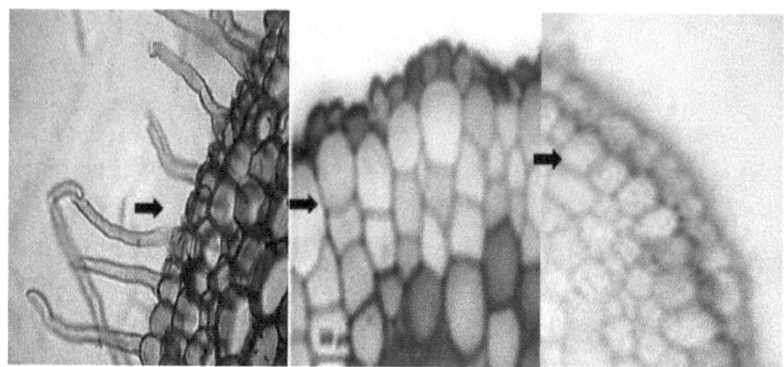

a: Piliferous bedb: Suberoid bedc: Suberous bed

Figure 2: Root lining tissues

2. Parenchyma

i- Definition:

It is a fundamental tissue within which the various specialised cell types will subsequently differentiate. The word parenchyma will therefore essentially refer to all the cells which do not appear to have any particular specialisation. Hence the tendency to consider it as a simple "filling" tissue. But in reality, parenchyma has functions no less important than those of other tissues. The functions of assimilation and reserves. The parenchymal cells have a pectocellulosic wall, are polygonal, of variable shape and size, with or without meats, and sometimes with lacunae.

ii- Type of parenchyma

Depending on the role of the parenchyma, a distinction is made between :

Assimilatory parenchyma are characterised by their "chlorophyll" plastids, and are located in the outer regions of stems and leaf blades. It is at this level, through photosynthesis, that energetic or plastic materials are produced (Figure

3).

The reserve parenchyma, located deeper down, accumulates the energy substances that the plant will use when the time is right. These materials accumulate in vacuoles or in the cytoplasm. These include starchy parenchyma in tubers and seeds, and cells with protein reserves (aleurone) or fat reserves in seeds (Figure 3).

Water-bearing parenchyma are made up of large cells with a highly developed vacuole, rich in water and often mucilage. This is a type of reserve parenchyma in which water is the compound stored.

Aeriferous parenchyma, found in aquatic plants, are varieties of lacunar tissue where the very large lacunae store air (Figure 3).

A distinction is made between :

The palisade parenchyma, a chlorophyllous parenchyma characterised by two or more rows of elongated, contiguous cells oriented perpendicular to the epidermis (Figure 3).

Lacunar parenchyma, in which the cells are not joined and are separated by lacunae, and meatus parenchyma, in which the cells form triangular spaces between three cells (Figure 3).

The meatus parenchyma is made up of several layers of polygonal cells. Between each three cells, there is a space called a meatus (Figure 3).

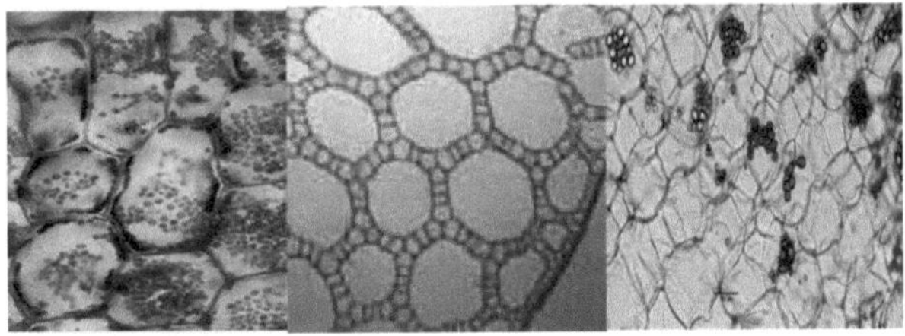

Assimilating parenchyma Aeriferous parenchyma Reserve parenchyma

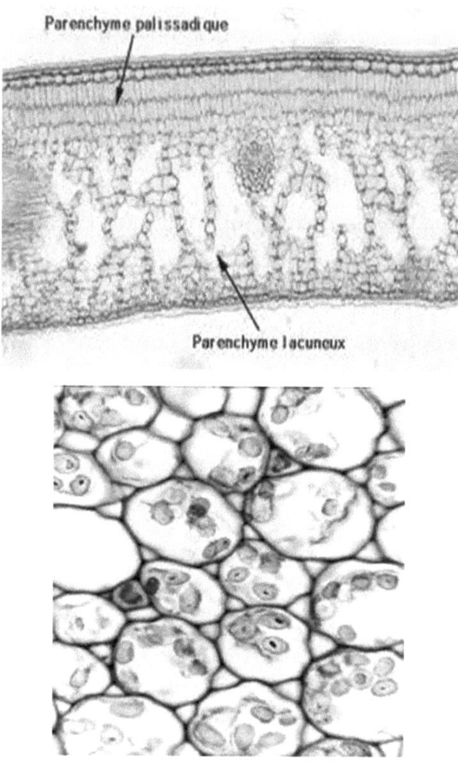

Parenchyma with meatus

Figure 3: Parenchyma

3. Support tissues

i- Definition:

The support apparatus is a conventional group of tissues whose mechanical role is to contribute to the strength of the plant. The strength, flexibility and elasticity of a stem result from the specific qualities and architecture of these support elements. All the supporting elements have a very thickened wall. Two groups can be distinguished according to the nature of their wall: collenchyma, formed by cells that have remained alive and whose walls are exclusively cellulose, and sclerenchyma, formed by cells that are generally dead and whose walls are more or less strongly lignified.

ii- Collenchyma :

It is the supporting tissue of young, aerial organs that have not yet completed their growth. Collenchyma is made up of living cells with walls thickened by a deposit of cellulose. In cross-section, they look different depending on how thickened their walls are (Figure 4).

• **The collenchyma is round** or **annular**. The wall is uniformly thickened on the inside; there are no meats.

• **Lamellar collenchyma**: the cells are separated by meats; only at the edges of the meats is the wall thickened.

• **The angular collenchyma:** the wall is only thickened at the corners the space usually occupied by the meats is filled by cellulose and the middle pectic lamella.

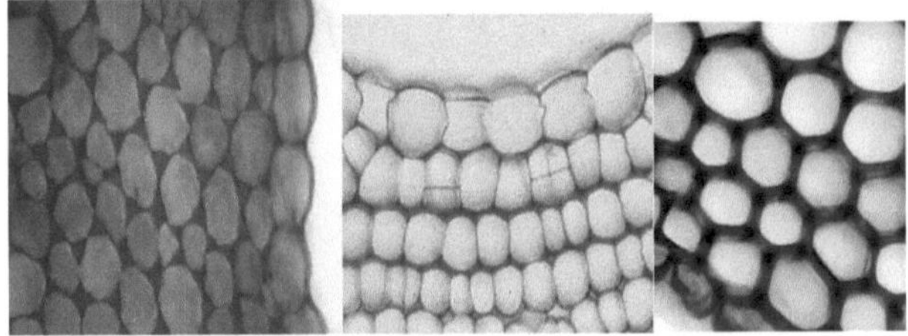

Angular collenchymaLamellar collenchymaRound collenchyma

Figure 4: Collenchyma

iii- The sclerenchyma

Sclerenchyma elements have thickened walls with varying degrees of lignification. A distinction is made according to their shape between sclerenchyma cells, sclerites and fibres (Figure 5).

• **The sclerotic cells** are roughly isodiametric. Their wall is entirely lignified and hollowed out by numerous canaliculi; it limits a lumen corresponding to the cell cavity, the contents of which have all disappeared. The sclerotic cells are dead cells.

• **Sclerites are** branched, usually large, sclerotic cells. Their presence is characteristic of certain tough organs.

• **Fibres are** elongated, spindle-shaped cells (in longitudinal section), with thick wall, more or less lignified, bordering a central cavity: the lumen, these are dead cells.

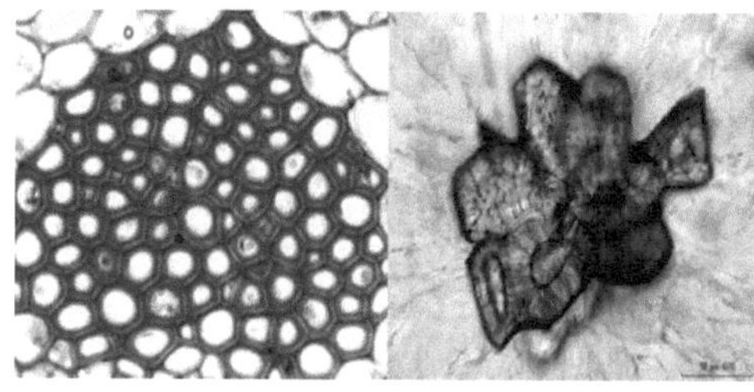

Sclerotic cells

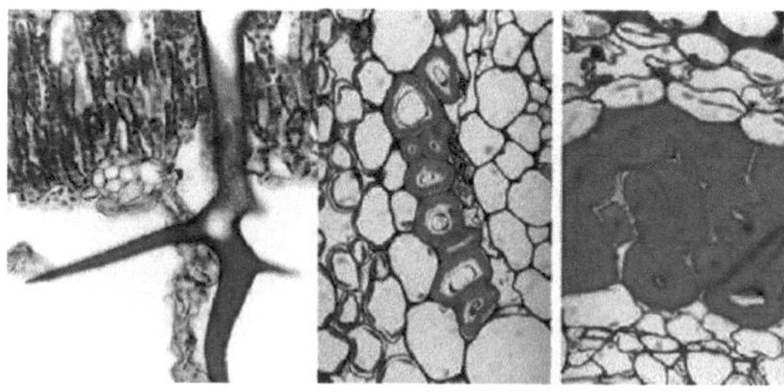

Scleritis Fibers

Figure 5 : The sclerenchyma

4. The conductive device

During the life of a plant, a double stream of liquids - sap - flows through its body. Raw sap is made up of water and mineral salts, drawn from the soil at root level, and circulates from bottom to top (ascending sap) to reach the various assimilating tissues. It is transported by the functional elements of the **xylem or vascular tissue**. Elaborated sap is a solution of organic matter, which circulates from top to bottom towards the tissues that use it. It is transported by the functional elements of the **phloem or sieve tissue.** Xylem and phloem together make up the conductive apparatus of plants or criblo-vascular tissue. The conducting apparatus is located in the central cylinder of the stem and root, and in the veins of the leaves (Figure 6).

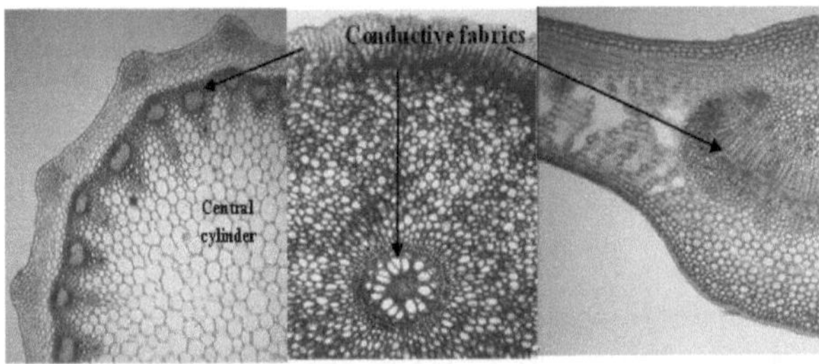

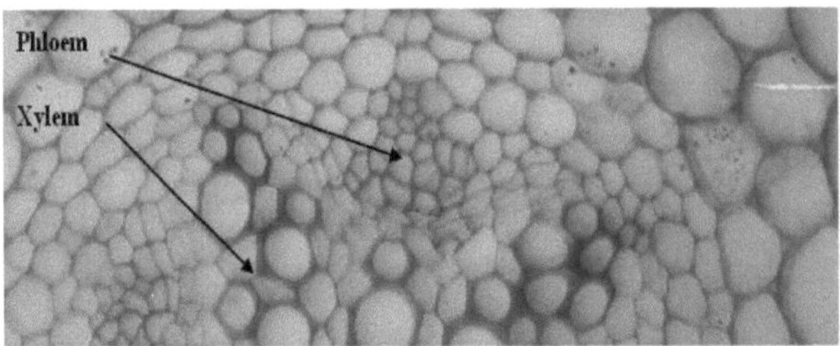

Figure 6: The conductive apparatus of plants

A. The xylem

It is a heterogeneous tissue, made up of Protoxylem and Metaxylem (Figure 7). **The protoxylem** consists of the very first differentiated woody elements before the organ has completed its elongation. **The metaxylem** appears in contact with the protoxylem after the organ has completed its elongation from cells that have acquired parenchymal characteristics.

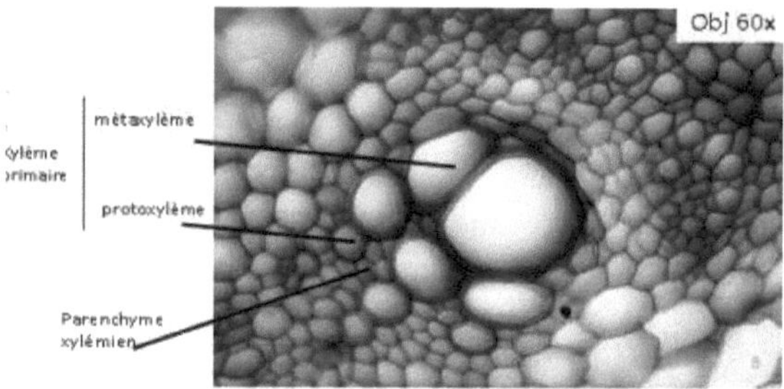

Figure 7: The xylem

The xylem is made up of conductive and non-conductive elements.

a. Conductive elements :

They are made up of elongated dead cells with lignin thickenings on the side walls separating cellulose patches. A distinction is made between imperfect vessels (tracheids) and perfect vessels (tracheae).

Tracheids: very elongated cells with tapering ends, lignified secondary walls and punctate terminal walls. They have transverse walls and therefore do not form continuous rows. Circulation takes place in chicane fashion. Depending on lignin impregnation, a distinction is made (Figure 8):

- **Ringed or spiral tracheids**: part of the protoxylem. Bevelled transverse partitions. The thickenings are arranged in a ring or spiral respectively.

- **Scalariform tracheids**: elements of the meta-xylem of ferns. A scalariform

24

tracheid has a polygonal cross-section with a bevelled end. Their arrangement is reminiscent of a ladder, hence their name. From one upright to the other, lignin thickenings form the rungs of the ladder, bordering cellulose patches.

• **Tracheids with areolate punctations**: part of the meta-xylem of gymnosperms, with bevelled ends, lignified walls and areolate punctations where the sap circulates.

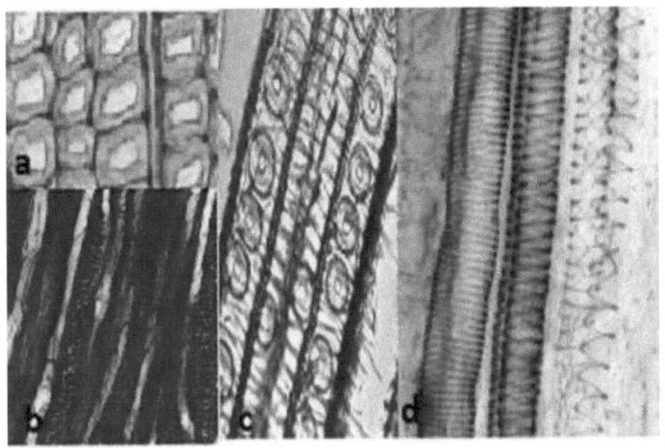

a: Transverse section, b : Longitudinal section, c : Areolated tracheids, d : Spiral tracheids

Figure 8: Tracheids

Tracheae (perfect vessels): These are made up of rows of aligned cells that fuse together end to end as they develop, with the transverse walls disappearing to form a kind of tube that can be several metres long. The longitudinal walls are thickened with lignin (Figure 9), and a distinction is made between :

• **Ringed and spiral vessels**: the thickening forms a ring-shaped or spiral ornamentation.

• **Striped vessels**: the ornamentation is in stripes

• **Reticulated vessels**: lignified wall network, transverse and perforated wall

• **Punctate vessels**: the widest and shortest, the transverse wall is resorbed.

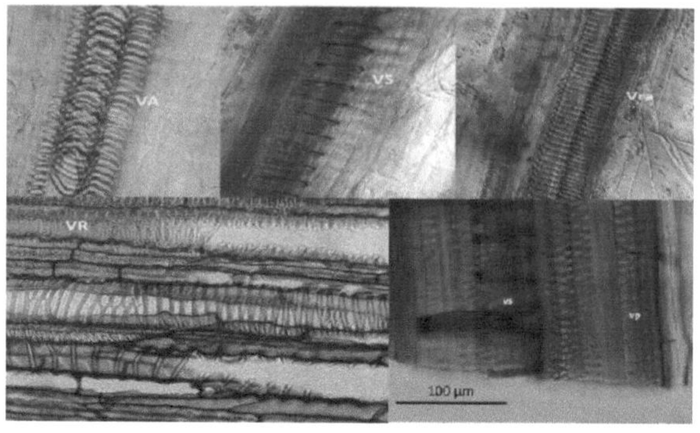

Figure 9: The vessels

b. Non-conductive elements

The woody parenchyma: made up of living cells with a pectocellulose or lignified wall, it plays a role in reserving or regulating the quantity of raw sap. Depending on its position, a distinction is made between vertical parenchyma (woody parenchyma proper) and horizontal parenchyma (medullary rays, inter ligneous) (Figure 10).

Fibres: elongated cells with a tapered end and a lignified wall, the contents of which have disappeared; they are punctate.

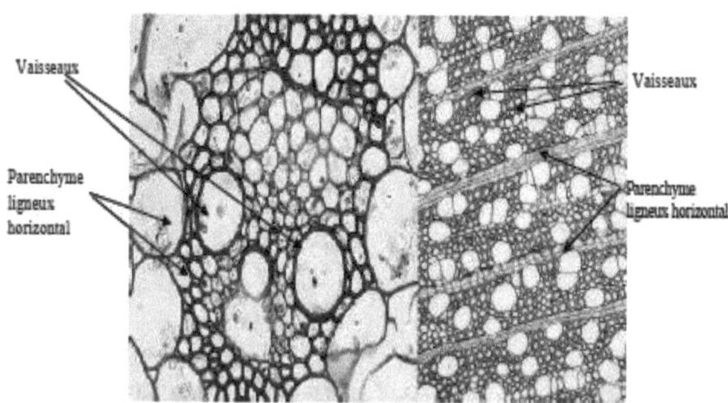

Figure 10: Woody parenchyma

B. Phloem: sieve tissue

This is a heterogeneous tissue (Figure 11). It is made up of sieve tubes, which are the conductors of the elaborated sap, and companion cells.

The sieve tubes: these are living elongated cells with a cellulose wall, the transverse walls of which are pierced by punctations that are uniformly distributed or grouped into sieve patches. The sieve tubes have a short lifespan, after which time the sieve plates are covered with a **"callus"** plug made up of a special substance called callose.

Companion cells (annex cells): These are formed by unequal division of the initial sieve tube cells. The companion cells take over after the sieve tubes lose their activity.

Parenchyma phloem : cells parenchymal cells common longitudinally elongated with an unscreened cellulose wall.

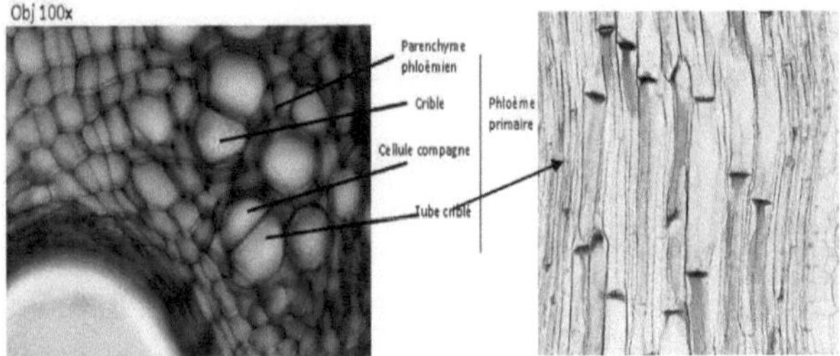

Figure 10: The phloem

II. Plant tissues of secondary origin

Secondary tissues are characteristic of Gymnosperms and Dicotyledons. They are the result of the functioning of secondary meristems (cambium). There are two types of secondary meristem (Figure 11).

The xylemophloem cambium: known as the vascular cambium or cambium for short. This is the most important cambium for the growth of a plant with secondary formations. Appearing within the conductive elements of the central cylinder, it produces, often in abundance, new conductive elements: wood (secondary xylem) towards the interior and liber (secondary phloem) towards the exterior.

The suberophellodermal or phellogenic cambium: this forms in the bark. The plant owes it two secondary tissues: the suber (cork), a covering tissue on the outside, and the phelloderma on the inside, a parenchymatous tissue.

Because of the periclinal mode of division of the cambium cells, the cells of the secondary tissues appear arranged in radial rows.

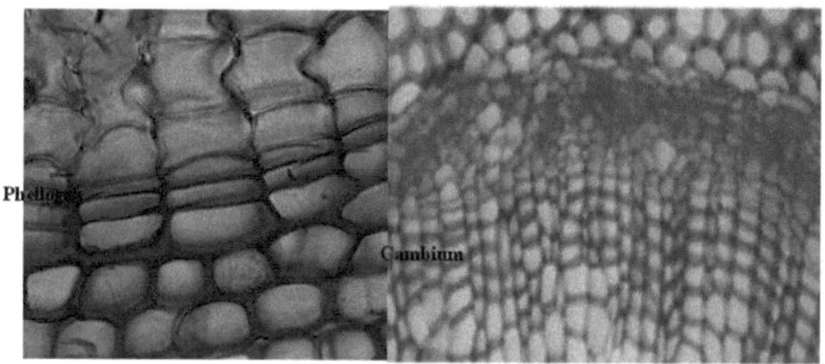

Figure 11: The cambiums (vascular on the right)

1. Wood (secondary xylem)

It grows inwards. It grows centripetally, synchronised with the seasons. It forms annual layers made up of two parts (Figure 12). Light-coloured spring wood with large cells, because the climatic conditions are favourable for growth, and darker, harder autumn wood with smaller cells. This wood makes up the majority of tree trunks.

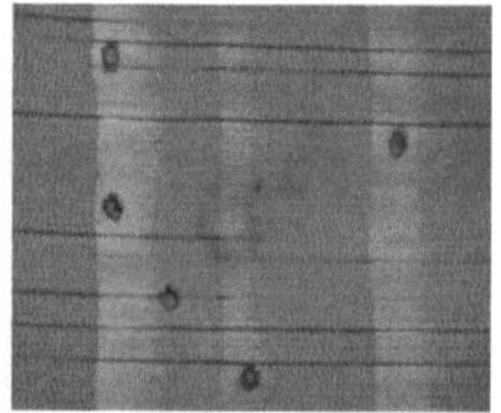

Figure 12: The annual layers of wood

Anatomical observation of the wood reveals a clear difference between that of gymnosperms and dicotyledonous angiosperms (Figure 13). In Gymnosperms, it has a uniform appearance and is made up of tracheids only, mainly areolate and devoid of vessels; it is said to be homoxylated. In Dicotyledons, it is made up of vessels and also tracheids, and is known as heteroxylated wood.

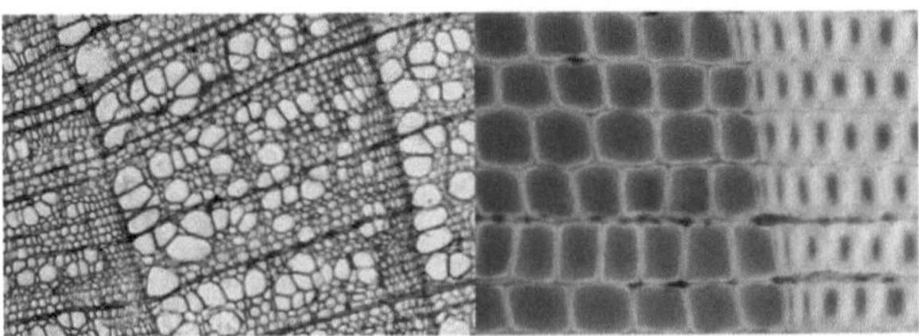

Figure 13: Types of wood (homoxylated on the right, heteroxylated on the left)

2. The liber

It develops inwards (Figure 14). The conductive elements of the liber are represented by the cells and sieve tubes. The non-conductive elements are formed by the vertical and horizontal liber parenchyma. Together with the wood parenchyma, the latter forms the medullary rays. These are radially oriented bands of cells that run from the medullary parenchyma to the cortical parenchyma and successively cross the wood and the bast from the inside out. The medullary rays are more or less wide, single, double or multiple.

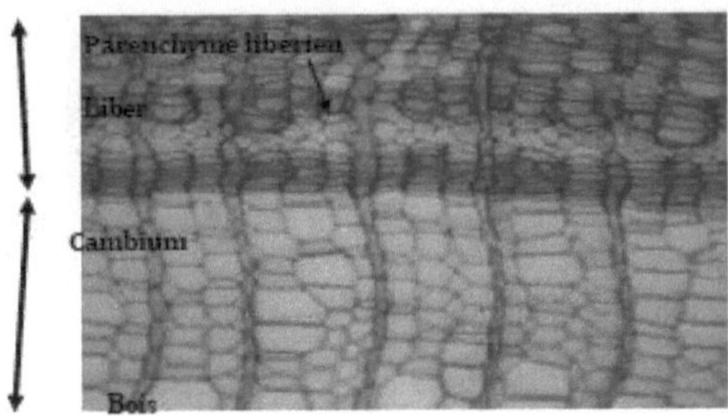

Figure 14: Secondary conductive fabric

3. The suber

Also known as cork, this is a secondary lining tissue, formed by several layers of rectangular, more or less thick-walled, suberified cells (Figure 15), with lenticels running through them.

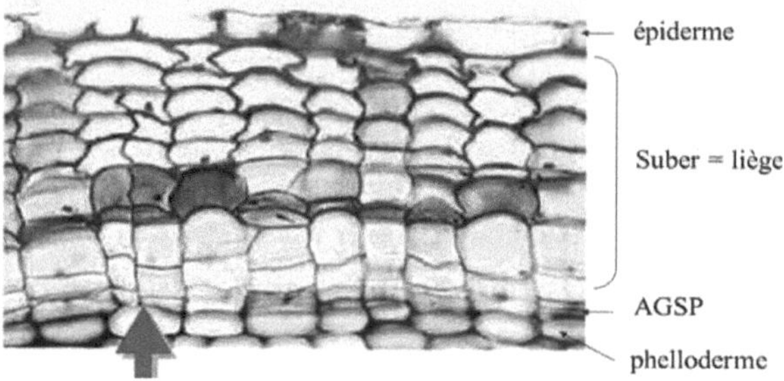

Figure 15: Suber

III. Secretory apparatus

1. Definition

The secretory apparatus is a group of cells or tissues in different locations, whose role is to produce, store and transport plant secretions. A distinction is made between :

- Secretory hairs

- Secretory pockets

- Secretory ducts

- Petrol cells

- The laticifers

- The nectaries

2. Secretory hairs

They are also known as glandular hairs. They are located in the epidermis of the aerial organs. They are made up of two parts, the head (containing the essential oils) and the feet (Figure 16). Depending on how the cells are organised, there are :

• Hairs with unicellular heads and feet

• Hairs with multicellular heads and unicellular feet

• Hairs with unicellular heads and multicellular feet

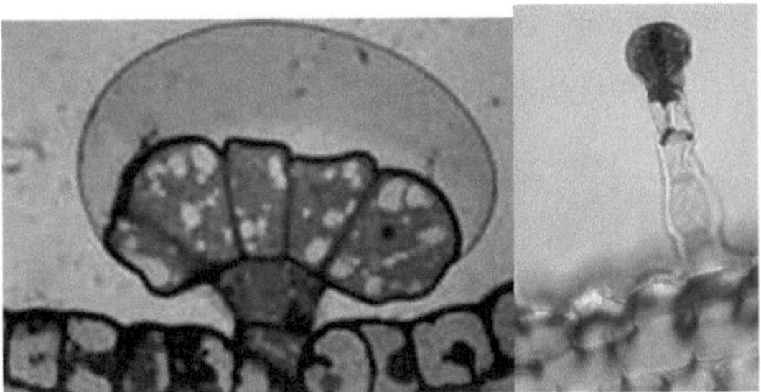

Figure 16: Secretory hairs

3. Secretory pockets

They are found inside the bark. It is a spheroid structure, formed by an essence-producing layer in the middle of a parenchymatous tissue. The inner layers can break to release the essence. This is known as a schizolysigenous pocket. If the essence is released into the pocket without destroying the cells, this is known as a schizogenic pocket (Figure 17). The pockets appear as translucent spots on the leaves.

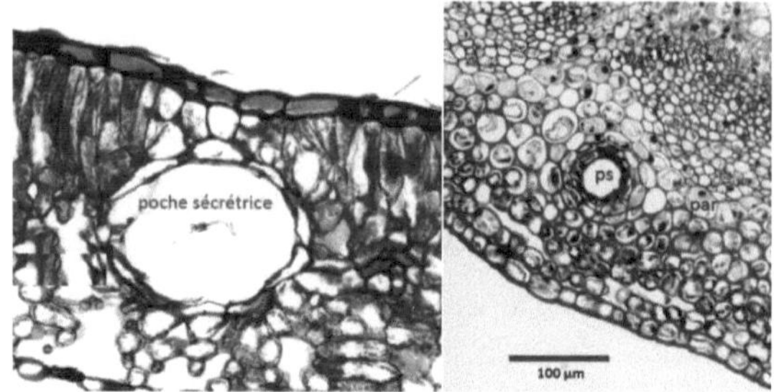

Schizolysigenous pouch Schizogenic pouchFigure 17: Secretory pouches

4. Secretory ducts

The lumen of a secretory duct is narrower than that of a pouch (Figure 18). They are found in various organs. Both in the bark and in the central cylinder.

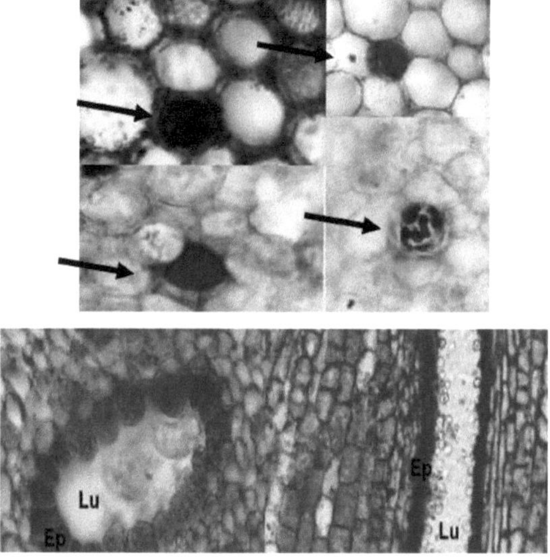

Lu: lumen of the duct, Ep: wall of the duct Figure 18: Secretory ducts

5. Fuel cells :

Generally, these are epidermal cells that store essences, like those found in rose petals (Figure 19). They appear with a dilated wall.

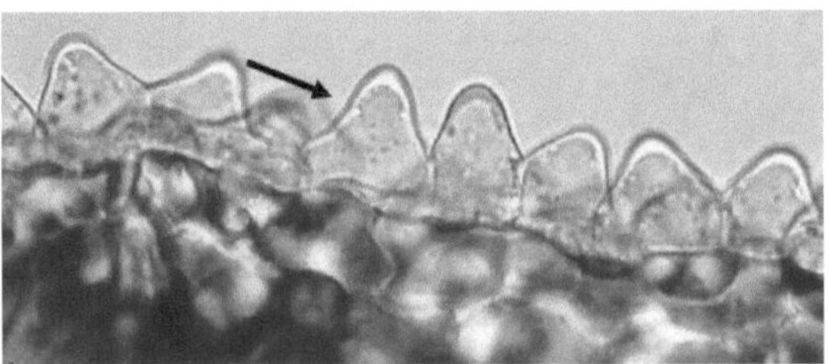

Figure 19: Petrol cells

6. The laticifers :

They run along the length of the organ, through the various tissues (Figure20). They secrete and transport latex, a milky substance containing alkaloids. The laticifers may be articulated or non-articulated (multinuclear), branched or unbranched.

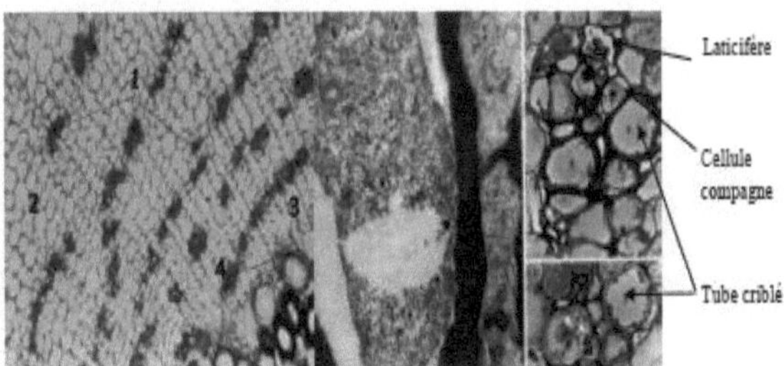

1: Laticifers, 2: Liber, 3: Wood, 4: Cambium

Figure 20: Laticifers

7. The nectaries

They are found on the flower. They secrete a sugary solution called nectar. A nectary consists of a set of secretory cells with a dense cytoplasm and a conductive tissue.

CHAPTER III
PLANT ANATOMY

1. Definition

The study of the organisation of tissues within the organ is known as plant anatomy. A plant organ is studied after histological sections have been taken, using different cross-sections.Anatomy consists of determining the position of tissues, their importance and their organisation. The description begins with the outline of the section and its symmetry, and the different regions that make it up. The root and stem have two regions, the bark on the outside and the central cylinder (CC) on the inside. The leaf is made up of a vein region and a blade region.Symmetry in the stem and root may be axial, bilateral or irregular. The contour may be circular or angular. In the leaf, symmetry is generally bilateral. There is a ventral or posterior side and a dorsal or anterior side.The study of plant anatomy begins with the identification of the different tissues, and ends with the production of a general diagram and a detailed drawing of the section observed.The term primary structure is given to the organ that has not completed the formation of secondary tissues. The vascular cambium, which generally appears early, is generally observed in organs with primary structures. The secondary structure is only found in Gymnosperms and Dicotyledons.

2. Anatomy of the root with primary structure

The root is generally circular in cross-section (Figure 16) and is characterised by:

The surface area of the central cylinder is smaller than that of the ecore.
The presence of coating tissues and reserve cortical parenchyma in the bark, and conductive tissues and medullary parenchyma in the CC.

The boundary between the CC and the bark is formed by two tissues which regulate exchanges (Figure 16): The endodermis: a single layer of contiguous, rectangular cells with a more or less thick wall and a mixed lipid thickening (lignin and suberin). In Monocotyledons, the deposit is U-shaped, hence the name U-shaped endodermis. In Dicotyledons, the deposit is found on both side walls, hence the name frame endodermis.The pericycle: a parenchymatous tissue formed by one or more layers of small parenchymatous cells.Conductive tissues alternate in the central cylinder. The xylem grows centripetally.

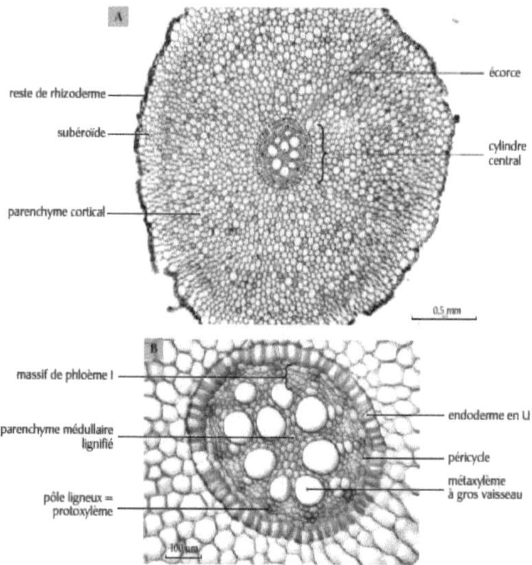

Figure 16: Anatomy of the primary root

3. Anatomy of the primary stem structure

The rod has an angular or circular cross-section (Figure 17) and is characterised by : The surface area of the central cylinder is greater than that of the spindle. The presence of coating tissue and cortical parenchyma in the bark, and conductive tissue and medullary parenchyma in the CC. The boundary between the CC and the bark is generally distinguished by a supporting tissue (sclerenchyma) forming a sheath, and also the conducting bundles in concentric circles. In Monocotyledons, a higher number of conductive bundles can be seen than in Dicotyledons (Figure 17B). A conducting bundle is formed by the superposition of xylem and phloem. The bundle is generally surrounded by supporting tissue. The xylem grows centripetally (Figure 17C).

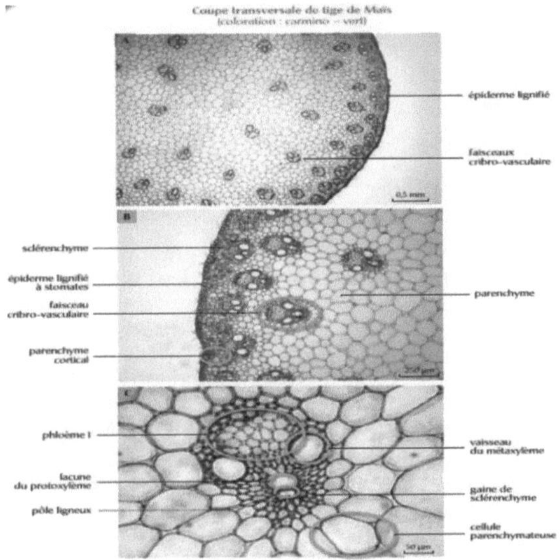

Figure 17: Anatomy of the primary stem

4. Anatomy of the leaf

The sheet has two regions:

Blade: formed from the upper surface to the lower surface by an upper epidermis, a mesophyll (leaf parenchyma) and a lower epidermis (Figure 18).

In Dicotyledons, the mesophyll is formed by two types of parenchyma, a palisading parenchyma on the upper surface and a lacunar parenchyma on the lower surface. This parenchyma is said to be heterogeneous and asymmetrical. In certain Dicotyledons, the lacunous parenchyma is found between two palisading parenchyma (upper and lower), the mesophyll is said to be symmetrical heterogeneous. In Monocotyledons, the mesophyll is formed by a single type of parenchyma and is said to be homogeneous symmetrical.

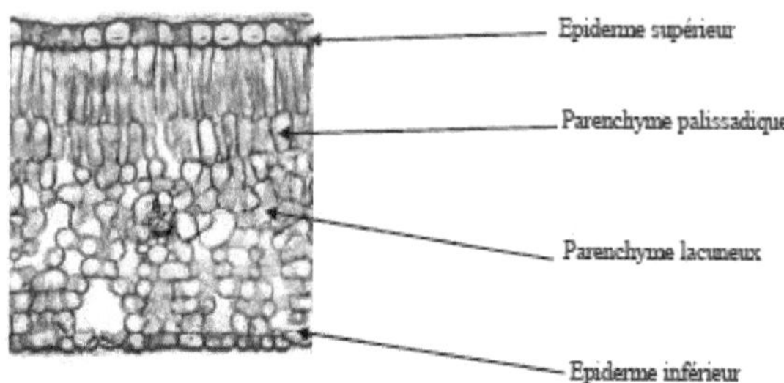

Figure 18: anatomy of the leaf blade

The vein: in the centre of the vein, there is a conductive bundle in the shape of an inverted arch. The phloem is on the underside and the xylem on the outside. Collenchyma is generally present on both sides of the vein (Figure 19).

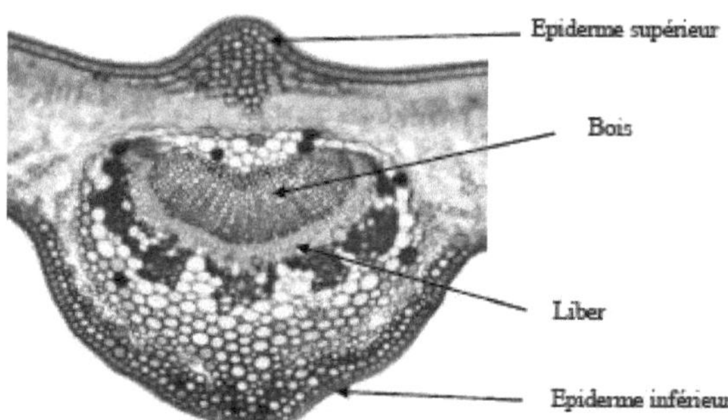

Figure 19: Anatomy of the vein

5. Anatomy of the root and secondary stem

Once established, the secondary structures invade the entire organ. The first secondary elements are the conducting elements, wood and bast. They take the place of the xylem and phloem respectively. The bundles formed are called a pachyte, which may be continuous or discontinuous. The suber is formed by rejecting the pre-existing conductive tissues towards the outside of the organ. These degenerate and disappear.In the root (Figure 20A), because the central cylinder is narrow, the pachyte completely invades the medullary parenchyma and crushes the primary conductive tissues, which generally disappear. The wider bark retains some of its cortical parenchyma. In the stem (Figure 20B), because the bark is narrow, the suber and phelloderma take up all the existing space. Whereas in the central cylinder, which is wider, part of the medullary parenchyma remains.

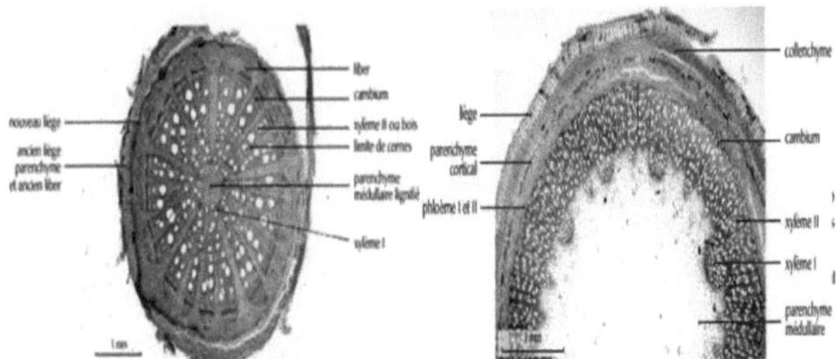

Figure 20: A root with secondary structure, B stem with secondary structure

REFERENCES

Bowes, B. G., & Mauseth, J. D. (2008). Plant structure: a colour guide. Manson.

Beck, C. B. (2010). An introduction to plant structure and development: plant anatomy for the twenty-first century. Cambridge University Press.

Cutler, D. F., Botha, T., & Stevenson, D. W. (2008). Plant anatomy. An applied approach. Malden, MA: Blackwell Publishing.

Deysson, G. (1965). Elements of vascular plant anatomy. Dickison, W. C. (2000) Integrative plant anatomy. Academic press.

Schweingruber, F. H., Börner, A., & Schulze, E. D. (2011). Atlas of Stem Anatomy in Herbs, Shrubs and Trees: Volume 1 (Vol. 1). Springer Science & Business Media.

I want morebooks!

Buy your books fast and straightforward online - at one of world's fastest growing online book stores! Environmentally sound due to Print-on-Demand technologies.

Buy your books online at
www.morebooks.shop

Kaufen Sie Ihre Bücher schnell und unkompliziert online – auf einer der am schnellsten wachsenden Buchhandelsplattformen weltweit! Dank Print-On-Demand umwelt- und ressourcenschonend produzi ert.

Bücher schneller online kaufen
www.morebooks.shop

info@omniscriptum.com
www.omniscriptum.com